Nuclear test Environment

Marshall Islands: Bikini Atoll, Enewetak Atoll, Rongelap Atoll, Utrōk Atoll

Author
Ostenfosh Burnns

Copyright Notice

First Printing: 2017.

ISBN: 978-1-912483-52-5

Publisher: Global Print Digital.
Arlington Row, Bibury, Cirencester GL7 5ND
Gloucester
United Kingdom.
Website: www.homeworkoffer.com

Table of Content

Introduction

Strategic Initiatives

The United States Department of Energy has recently implemented a series of strategic initiatives to address long-term radiological surveillance needs at former U.S. nuclear test sites in the Marshall Islands. The plan is to engage local atoll communities in developing shared responsibilities for implementing radiation surveillance monitoring programs for resettled and resettling populations in the northern Marshall Islands.

Using the pooled resources of the United States Department of Energy and local atoll governments, individual radiological surveillance programs have been developed in whole body counting and plutonium urinalysis. These programs are used to accurately track and assess doses delivered to Marshall Islanders from exposure to residual fallout contamination in the local environment.

The key residual fallout radionuclides of radiological concern include fission products, such as cesium-137 and strontium-90, and long-lived alpha emitting radionuclides including plutonium-239, plutonium-240 and americium-241. Permanent whole body counting facilities have been established at three separate locations in the Marshall Islands (Figure 1). These facilities are operated and maintained by Marshallese technicians with scientists from the Lawrence Livermore National Laboratory providing on-going technical support services.

The concentration of cesium-137 in soils from the northern Marshall Islands is significantly elevated over that expected from global fallout deposition and may enter the body of local residents through ingestion of locally grown foods. Whole-body counting provides a direct measure of internally deposited cesium-137 and is a very reliable method for assessing the internal dose contribution from ingestion of cesium-137.

We have also developed a state-of-the-art measurement technology in support of the Marshall Islands plutonium urinalysis (bioassay) program. Bioassay samples are collected by locally trained technicians under controlled conditions and returned to the United States for analysis of plutonium isotopes by Accelerator Mass Spectrometry (AMS). High-quality bioassay measurements based on AMS are providing more reliable and accurate baseline measurements, and could potentially be used track and assess intakes of plutonium associated with resettlement activities in the northern Marshall Islands.

Site specific environmental surveys are also conducted to determine the fate and transport of fallout radionuclides in the environment or simply to verify the effects of cleanup programs. The general aim of the environmental studies program is to develop fundamental scientific data on the behavior of key radionuclides in the environment. The data and information developed from these studies will ultimately be used to develop more reliable dose assessments for resettlement taking into account future change in radiological conditions. This information is essential in helping determine the most appropriate measures for cleanup and in assessing the impacts of changes in life-style, diet, and land-use on radionuclide uptake and dose.

Together, the individual and environmental radiological surveillance programs in the Marshall Islands are helping meet the informational needs of the United States Department of Energy and the

Republic of the Marshall Islands. Our mission is to provide high quality measurement data and reliable dose assessments, and to build a strong technical and scientific foundation to help sustain resettlement of affected atolls. Perhaps most importantly, the recently established individual radiological surveillance programs provide atoll population groups with an unprecedented level of radiation protection monitoring where for the first time, local resources are being made able to monitor resettled and resettling populations on a more permanent basis.

This web site provides an overview of the individual radiological surveillance programs currently being employed in the Marshall Islands along with a full disclosure of verified measurement data. A feature of this web site is a provision whereby users are able to calculate and track radiation doses delivered to program volunteers based on de-identified measurement data.

Brief History of Nuclear Testing in the Marshall Islands

Joint Task Force

Immediately after World War II, the United States created a Joint Task Force to develop a nuclear weapons testing program. Planners examined a number of possible locations in the Atlantic Ocean, the Caribbean, and the Central Pacific but decided that coral atolls in the northern Marshall Islands offered the best advantages of stable weather conditions, fewest inhabitants to relocate, and isolation with hundreds of miles of open-ocean to the west where trade winds were likely to disperse radioactive fallout.

During the period between 1945 and 1958, a total of 67 nuclear tests were conducted on Bikini and Enewetak Atolls and adjacent regions within the Republic of the Marshall Islands. The most significant contaminating event was the Castle Bravo test conducted on March 1, 1954 (Figure 2). Bravo was an experimental thermonuclear device with an estimated explosive yield of 15 MT (USDOE, 2000) and led to widespread fallout contamination over inhabited islands of Rongelap and Utrōk Atolls, as well as other atolls to the east of Bikini.

Today, the U.S. Department of Energy (U.S. DOE) through the Office of Health Studies continues to provide environmental monitoring, healthcare, and medical services on the affected atolls.

Key directives of the Marshall Islands Dose Assessment and Radioecology Program conducted at the Lawrence Livermore National Laboratory are (1) to provide technical support services and oversight in establishing radiological surveillance monitoring programs for resettled and resettling populations in the northern Marshall Islands; (2) to develop comprehensive assessments of current (and potential changing) radiological conditions on the islands; and (3) provide recommendations for remediation of contaminated sites and verify the effects of any actions taken.

Nuclear Testing in Marshall Islands

Immediately after the end of World War II the United States sought out a location where it could test and develop its newly proven and developed Nuclear Arsenal. The location decided upon would be a series of Atolls in the Marshall Islands in the South Pacific, most notably Bikini, seen in Fig. 1, and Enewetak Atoll. The location was approved by the United Nations as the Strategic Trust Territory. In these Pacific islands, as many as 105 tests occurred, totaling around 210 megatons. Among these tests were the first Hydrogen or

Thermonuclear weapons, much more powerful than the fission bombs from beforehand. Many unique tests were carried out, including flying B-17 drones over zero point at detonation to see the damage and testing weapons on a fleet of decommissioned warships. Tests were conducted on land and in water, causing mass amounts of radioactive fallout spreading to the 2,000+ islands that the Marshall Islands consist of and destruction of entire islands and coral reefs at ground zero. Many unforeseen effects from the radiation on the islands have left residents with health problems and long lasting effects on their ecosystem.

Castle Bravo

Castle Bravo, the first dry thermonuclear bomb and probably the most notable of all the tests, was subject to massive error and miscalculation by scientists. The device, originally planned to be in the range of 4 to 8 megatons was measured at 15 megatons. The basic idea of a Hydrogen bomb is to utilise fusion through fission instead of just fission. A conventional fission bomb is used as the trigger to compress the secondary part of the bomb, consisting of fusion fuel, 40% Lithium-6 deuteride and Lithium-7 deuteride, a dry fuel. When talking about a dry Hydrogen bomb we are talking about a solid and not a liquid or wet fuel.

The use of dry fuel is favored due to the fact that Lithium-6 deuteride is much easier to weaponize than deuterium and tritium gases and liquids, which in turn allows for a dramatic cut in weight of the device, allowing many more delivery capabilities. The reason the yield of the explosion was so high was due to the use of the dry fuel and its unknown capabilities, leading to miscalculations by the designers from Los Alamos Laboratories. This error was due to the fact that the Lithium-7 deuteride was expected to be inert when accepting a neutron and becoming a decaying isotope, Lithium-8.

This decay indeed occurred, but into a high energy alpha particle, a neutron and a tritium nucleus, which in turn lead to be a part of the drive. This caused fusion between the tritium and the deuteride and lead to increased yield that was not expected. we can see the fireball of 4 miles in diameter within the first second of detonation and contaminating 7,000 miles of the Pacific Ocean. The fallout from the explosion was much larger than expected and affected many of the nearby islands and the villagers in them.

Effects on Environment and People

The US Government had relocated all residents of the Bikini and Enewetak atolls to shield them from direct effects of the nuclear fallout. However, fallout from the testing was vast and reached locations the government did not expect to be touched by the testing. Most effects that occurred came through the Castle Bravo test due to the fact that it was so much larger than expected, particularly in the northern atolls and on Rongelap atoll.

Residents of Rongelap islands had experienced symptoms of itchiness, vomiting and fatigue, common to radiation sickness. As seen in Fig. 3, people living in these northern atolls have been measured to experience between 200 and 1,000 mGys, while the average US resident experiences 1 mGy. Through calculating these numbers researchers have projected that there will be an estimated 170 radiation caused cancer for people in the Marshall Islands.

Most of these cancers will be of thyroid cancer or some form of leukemia. These numbers vary depending on age, as children are much more susceptible to radiation in their thyroid than adults. Not only did the testing effect the people, but also the ecosystem in which they lived in. With the vast amount of testing that occurred in the water the coral reefs of the islands were virtually decimated and the habitat in which they sustained.

However sixty three years later, marine life in Bikini Atoll flourishes with coral reefs growing and fish plentiful. While there are no exact numbers on the number of species present today, we know of 126 coral species that existed before the tests and 42 that have not grown back. This demonstrates the biodiversity of the ecosystem and coral. While the coral has regrown, the islands remain unlivable to this day.

Marshall Islanders Reflect on a Dark Legacy of Nuclear Testing

The Republic of the Marshall Islands have been making ripples in global news lately. Fresh off a strong gathering at COP 21 in Paris, where the Honorable Tony de Brum (ex. Minister of Foreign Affairs in the Marshall Islands) rallied with other global leaders to advocate for more stringent policies for combating climate change, the small atoll nation of approximately 70,000 inhabitants also recently welcomed President Hilda Heine as their new Head of State.

Not only is President Heine the first female president in the country's history, she is also the first female president of ANY independent Pacific Island nation.

This month, the people of the Marshall Islands are shifting their focus on another critical issue. On March 1st, the Marshallese took time to observe Nuclear Victims Remembrance Day, which kicks off a month-long remembrance of a dark legacy that they hope the world will never forget. On this day, in 1954, a hydrogen bomb (code named "Castle Bravo") was dropped on Bikini Atoll in the Marshall Islands–a bomb that was hundreds of times stronger than the atomic bombs dropped on Hiroshima and Nagasaki.

This was part of the nuclear weapons testing program conducted by the US in the Pacific Islands and has led to numerous documented cases of illnesses related to radiation. For the Marshallese, this is a painful memory that still resonates strongly in the stories of elders and the images and videos of *Operation Castle* that have since been released. This is why every year, the month of March is dedicated to making sure that people in the Marshall Islands and around the world do not forget about the consequences of war and neocolonialism. This is especially cogent in light of the current intentions by the US Government to turn the island of Pagan into a live-fire training site.

One notable effort to raise awareness is the following collaboration between acclaimed Marshallese poet, Kathy Jetnil-Kijiner, and several students at the College of the Marshall Islands (CMI) Media Club. This group of young passionate storytellers – with support from the Pacific islands Climate Education Partnership (PCEP) – was able to create a powerful video to capture the emotional and physical toll of this legacy. Having only minimal media/journalism experience and no proper training, what the students of the CMI Media Club were able to produce is a testament to their talent and dedication for telling the story of their island home and people

The Legacy of U.S. Nuclear Testing in the Marshall Islands

The radiological legacy of U.S. nuclear weapons testing in the Marshall Islands remains to this day and will persist for many years to come. The most severe impacts were visited upon the people of the Rongelap Atoll in 1954 following a very large thermonuclear explosion which deposited life-threatening quantities of radioactive fallout on their homeland.

They received more than three times the estimated external dose than to the most heavily exposed people living near the Chernobyl nuclear accident in 1986. It took more than two days before the Rongelap people were evacuated after the explosion. Many suffered from tissue destructive effects, such as burns, and subsequently from latent radiation-induced diseases.

In 1957, they were returned to their homeland even though officials and scientists working for the U.S. Atomic Energy Commission (AEC) determined that radiation doses would significantly exceed those allowed for citizens of the United States. The desire to study humans living in a radiation-contaminated environment appeared to be a major element of this decision. A scientist in a previously secret transcript of a meeting where they decided to return the Rongelap people to their atoll stated an island contaminated by the 1954 H-Bomb tests was "by far the most contaminated place in the world." He further concluded that,

"It would very interesting to go back and get good environmental data... so as to get a measure of the

human uptake, when people live in a contaminated environment...Now, data of this type has never been available. ...While it is true that these people do not live, I would say, the way Westerners so, civilized people, it is nevertheless also true that they are more like us than the mice."

By 1985, the people of Rongelap fled their atoll, after determining that the levels of contamination were comparable to the Bikini atoll where numerous nuclear devices were detonated. The Bikini people were re-settled in 1969 but had to evacuate their homes in1978 after radiation exposures were found to be excessive. The Rongelap people fled for good reason. In 1982, a policy was secretly established by the energy department during the closing phase of negotiations between the United States and the nascent Republic of the Marshall Islands over the Compact of Free Association to eliminate radiation protection standards, so as to not interfere with the potential resumption of weapons testing. This resulted in a sudden and alarming increase in radiation doses to the Rongelap people eating local food.

These circumstances were subsequently uncovered in 1991 by the U.S. Senate Committee on Governmental Affairs. As a result, the U.S. Congress terminated DOE's nuclear test readiness program in the Pacific and in 1992 the U.S. Departments of Energy and Interior entered into an agreement with the Republic of the Marshall Islands and the Local Rongelap Government that re-established radiation protection standards as a major element for the re-settlement of Rongelap. Apparently, this was not done for the southern islands of the atoll where local food is obtained. Despite the long and unfortunate aftermath of nuclear testing in the Marshalls, it appears that this critical element of safety was lost in the shuffle.

As it now stands, if forced to return to their homeland the Rongelap people could receive radiation doses about 10 times greater than allowed for the public in the United States.

Until the U. S. Government can assure that steps to mitigate doses to the same levels that are protective of American people are demonstrated, efforts to force the Rongelap people back to the home by Members of the U.S. Congress and the Obama Administration is unjustified and unfairly places the burden of protection on the Rongelap people. It appears that DOE and Interior have quietly crept away from the 1992 agreement, without verifying that its terms and conditions to allow for safe habitability will be met.

Over the past 20 years, the U.S. Congress has enacted legislation to compensate to residents living near DOE's Nevada Test Site uranium miners, nuclear weapons workers, and military personnel for radiation-related illnesses. These laws provide for a greater benefit of the doubt than for the people of the Marshall Islands where 66 nuclear weapons were exploded in the open air.

In 2005, the National Cancer Institute reported that that the risk of contracting cancer for those exposed to fallout was greater than one in three.

The people of the Marshall Islands had their homeland and health sacrificed for the national security interests of the United States. The Obama Administration and the U.S. Congress should promptly correct this injustice.

Bikini Atoll

Summary Information

Bikini Atoll is one of two sites in the Republic of the Marshall Islands used by the United States for testing of atmospheric nuclear weapons (1946-58).

The nuclear test program on Bikini Atoll produced close-in fallout contamination over much the atoll forming a continuous source-term for remobilization of fallout radionuclides into the marine and terrestrial environment. Of importance, long-lived fallout radionuclides were assimilated to different levels into plants, animals and other organisms used for human consumption.

Residual levels of fallout radioactivity on the atoll have been well characterized, especially for the main residence islands of Bikini and Eneu, and together with information developed from individual radiological surveillance monitoring programs provide a good understanding of the important pathways for human exposure to residual falout contamination. Today, the key residual fallout radionuclides of potential radiological concern include cesium-137 and strontium-90 and, to a lesser extent, plutonium isotopes and americium-241.

The most important pathway for human exposure to residual fallout contmaination is ingestion of cesium-137 contained in locally grown food crop products such as coconuts, Pandanus, and breadfruit.

The predicted annual effective dose for resettlement of Bikini Island in 1999 without any form of remediation ranged from 15 mSv (1500 mrem) for a local foods only diet to about 4.0 mSv (400 mrem) where imported foods were made available. The compares with the average natural background dose in the Marshall Islands of about 1.5 mSv (150 mrem). The Republic of the Marshall Islands has adopted a cleanup standard of 0.15 mSv or 15 mrem per year above background.

In order to reduce the ingestion dose, researchers from the Lawrence Livermore National Laboratory have evaluated several methods for blocking the uptake of cesium-137 into plants and food-crop products. The most effective and practical method for reducing the uptake of cesium-137 into locally grown foods is to treat agricultural areas with potassium fertilizer (KCl). The addition of potassium has the added benefit of increasing the growth rate and productivity of some food crops.

Assuming imported foods are available and a resettlement date of 1999, the combined effects of treating the agricultural areas with potassium fertilizer and removing of the top 40 cm of contaminated soil in the village and housing area, reduced the estimated population average annual effective dose on Bikini Island from about 4 mSv per year to 0.41 mSv per year. Cesium-137 contributes about 90% of the estimated total annual effective dose via the ingestion pathway. External exposure accounts for about 10% of the total annual effective dose. In addition, natural environmental processes are helping reduce the amount of cesium-137 taken up into plants and other food-crop products.

This accelerated environmental-loss rate of cesium-137 reduces the cesium-137 burden in plants with an effective half-life of about 8.5 years compared with its radiological half-life of 30 years or about 3.5 times faster than predicted in existing dose assessments. Based on this new information, the population average annual effective dose on Bikini Atoll in 2010 after employing prescribed cleanup options and using a mixed diet with imported foods will be around 0.17 mSv (17 mrem) or very close to the self-imposed cleanup standard of 0.15 mSv (15 mrem) as adopted by the Marshall Islands Nuclear Claims Tribunal.

Upon implementation of a Bikini resettlement program, we recommend that a post-resettlement radiological surveillance program be established on Bikini Island based on whole body counting and, if required, plutonium urine bioassay. In this way the U.S. Department of Energy in cooperation with the Bikini Council can help ensure that radiological conditions remain at or below applicable safety standards.

A number of whole body counting facilities have already been established in other parts of the Marshall Islands and are operated by trained Marshallese technicians. Whole body counting provides a direct measure of the amount of cesium-137 in a person's body and does not rely on modeled assumptions associated with predictive dose assessments. An individual radiation protection-monitoring program on Bikini will also allow potential 'high-end' doses to be accurately tracked and assessed.

People and Events on Bikini Atoll

Bikini Atoll is one of two sites in the northern Marshall Islands used by the United States for testing of atmospheric nuclear weapons. Twenty-three nuclear devices were detonated on Bikini Atoll between 1946 and 1958 with a combined fission yield of 42.2 Megaton (Mt) (UNSCEAR, 2000). An additional forty-three atmospheric nuclear tests were conducted on Enewetak Atoll about 300 km to the west of Bikini Atoll.

The most significant contaminating event in the Marshall Islands nuclear test campaign and the highest yield atmospheric nuclear test ever conducted by the United States involved the detonation of a high-energy thermonuclear on Bikini Atoll on 1 March of 1954. This ground-surface test was code named Bravo and had an estimated explosive yield of 15 Mt (USDOE, 2000).

It is estimated that about 50% of the fission yield associated with near-surface nuclear detonations was deposited on a local or regional scale (Hamilton 2004; UNSCEAR, 2000). The remainder of the debris from near-surface denotations and all the debris from high altitude airbursts entered the global environment producing a worldwide pattern of global-fallout deposition.

Prior to Bravo, little consideration was given to the potential health and ecological impacts of fallout contamination beyond the immediate 'boundaries' of the test sites. However, regional fallout from the Bravo test unexpectedly caused widespread fallout contamination over Bikini Atoll and forced the evacuation of Marshallese people living on Rongelap and Utrōk Atolls (Cronkite et al., 1955).

Environmental Characterization of Bikini Atoll

Through the early 1980s, scientists from the Lawrence Livermore National Laboratory developed an extensive database of environmental measurements for Bikini Atoll, especially for soils and vegetation growing on Bikini and Eneu Islands. These detailed monitoring surveys were used to develop predictive dose assessments of exposure of hypothetical resident populations to residual fallout contamination in the marine and terrestrial environments. Wherever possible, the monitoring surveys involved direct measurements of radionuclide concentrations in soil and associated food-crop products, as well as air, water, fish, and resident marine organisms.

These data and information were essential in helping identify the key radionuclides and radiological exposure pathways in the Marshall Islands, and in assisting the U.S. Department of Energy and the Bikini Council in making more informed decisions about resettlement of the atoll. During this period, predictive dose assessments for both Bikini and Enewetak Atolls clearly indicated that the most significant pathway for human exposure to residual fallout contamination in the Marshall Islands was ingestion of cesium-137 contained in locally grown root crops such as coconut, breadfruit, and Pandanus (Robison et al., 1980; 1982).

Remediation Options

One key factor that helps explain why cesium-137 plays such a important role in contributing to radiation exposure in the Marshall Islands is that coral soils are known to contain little or no clay material and very low concentrations of naturally occurring potassium an alkaline earth element that shares similar properties with cesium. These conditions result in increased uptake of cesium-137 from soil and incorporation into plants relative to the rate of cesium-137 uptake from continental soils.

Consequently, the significance of dietary intakes of cesium-137 from eating locally grown foods was initially overlooked because early models in radioecology were based on continental type soils and exposure conditions. Knowledge of the unique behavior of cesium-137 in potassium-poor coral soil environments has also been instrumental in helping guide remediation experiments designed to reduce the dose delivered to resettled or resettling populations.

The first of a series of long-term field experiments was established on Bikini Island during the late 1980s to evaluate potential remediation techniques to reduce the uptake of cesium-137 into plants (Robison and

Stone, 1998). Based on these experiments, the most effective and practical method for reducing the uptake of cesium-137 into food crop products was to treat agricultural areas with potassium fertilizer (KCl). The addition of potassium had the added benefit of increasing the growth rate and productivity of some food crops with essentially no adverse environmental impacts.

One alternative is to excavate the top 30 to 40 cm of soil, but this type of remedial process would be much more expensive to implement over a large area. Furthermore, excavation carries away all the soil organic matter needed to maintain the water retention capacity of coral soils and supply essential nutrients to support plant growth. Soil excavation also necessitates a very long-term commitment to rebuild the soil and revegetate the land.

Large-scale field experiments on Bikini Island have been used to optimize the required amount and application rates of potassium (Figure 3). The results from these experiments show that a single application of 2000 kg per ha of potassium can be effective in reducing the cesium-137 uptake in coconut meat (and juice) to about 5% to 10% of the pretreatment level.

Multiple applications (over several months) of the same total amount of potassium produce even better and more consistent results. Moreover, the concentration of cesium-137 in the coconuts following remediation remains low for an extended period of time, so the need for continuous effort and retention of scientific and technical expertise is minimized (Robison et al., 2004). In fact, the use of potassium was adopted by the Rongelap Atoll Local Government (RALGOV) as part of a combined option for rehabilitation and cleanup of Rongelap Island (Rongelap Atoll).

The combined option calls for (1) the treatment of agricultural areas of the island with potassium fertilizer to reduce the uptake of cesium-137 into plants, and (2) the replacement of contaminated surface soil around the village and housing areas with crushed-coral fill in order to help minimize external exporsure rates in areas where people spend most of their time. This same type of approach would be applicable to reducing dose rates on Bikini Island.

Prospects for Resettlement of Bikini Atoll

Radiological for Bikini Island have been developed from samples collected and analyzed as part of a continuing environmental monitoring program on Bikini Atoll (1975-1994). These on-going studies have concentrated on assessing the uptake and remediation of cesium-137 in terrestrial plants because cesium-137 delivers, by far, the largest fraction of the radiation dose Using empirical data from annual or semi-annual monitoring surveys of selected trees on Bikini and Eneu Islands, we have recently demonstrated that the environmental half-life of cesium-137 is more important than radiological decay in controlling the fate and distribution of cesium-137 in coral soils (Robison et al., 2003).

For example, the estimated effective half-life of cesium-137 on Bikini, Enewetak, and Rongelap Atolls is around 8 to 9.8 years (95% confidence) compared with its radiological half-life of 30 years. These findings suggest that predictive dose assessments based on historical radiological decay-corrected measurement data may not be applicable to current or future radiological conditions on the atoll. With this knowledge, Livermore scientists have directed their scientific studies on Bikini Island towards quantifying the rates of environmental loss of cesium-137 using lysimeters and measuring the amount of cesium-137 washed out of the soil into the underlying groundwater.

Lawrence Livermore National Laboratory researchers have also developed an understanding of the residence time of the fresh water lens (and associated contaminants) on the island. Preliminary data also suggests that labile soil cesium-137 is slowly being incorporated into more resistant mineral phases within the soil and, through aging effects, may be becoming less available for soil-to-plant transfer. The lysimeter and ground water sampling program was terminated in December 2006 so the results from these studies should become available over the next year.

Although the mechanisms involved and the impact of adding potassium on soil cesium-137 mobility is not fully understood, the data and information stemming from the research and monitoring program on Bikini Island will enable more accurate dose predictions to be developed for various resettlement and cleanup scenarios on the island (and on coral atoll environments, in general).

Applying a mean effective cesium-137 half-life of 8.5 years for the data developed for the 1999 Bikini dose assessment (Robison et al., 1997a), the predicted population average effective dose for resettlement of Bikini in 2010, where imported foods are available, is conservatively estimated to be about 0.17 mSv per year (17 mrem per year) or very close to the self-imposed cleanup standard of 0.15 mSv per year adopted by the Republic of the Marshall Islands Nuclear Claims Tribunal. With this understanding and the fact that exposure conditions on Bikini are improving at an accelerated rate, early resettlement of Bikini Atoll may become much more plausible and cost effective.

Enewetak Atoll

People and Events on Enewetak Atoll

After an initial series of nuclear tests on Bikini Atoll in 1946, local inhabitants of Enewetak Atoll were relocated to a new home on Ujelang Atoll in December 1947 in preparation for scheduling of the first series of nuclear tests on Enewetak. Operation Sandstone commenced during April of 1948 and included 3 tests atop of 60 m high steel towers located separately on the islands of Enjebi, Aomen, and Runit.

An additional 4 near-surface tests were conducted on towers as part of Operation Greenhouse during 1951. Operation Ivy, in 1952, set the stage for the first test of a large thermonuclear device. The Mike thermonuclear blast of 31 October of 1952 had an explosive yield of 10.4 Mt (USDOE, 2000) vaporizing the island of Elugelab and leaving behind a deep crater about 1 km in diameter. Early analysis of Mike fallout debris showed the presence of two new isotopes of plutonium, plutonium-244 (244Pu) and plutonium-246 (246Pu), and lead to the discovery of the new heavy elements, Einsteinum and Fermium. Operation Castle involved a single nuclear test on Enewetak in 1954 and 5 high-yield tests on Bikini. A total of 11 nuclear tests were also conducted on Enewetak in 1956 as part of Operation Redwing including an air burst from a balloon located overwater.

In 1958, the United States anticipated the acceptance of a call for suspension of atmospheric nuclear testing and assembled a large number of devices for testing before the moratorium came into effect. From April through August 1958, 22 near-surface nuclear denotations were conducted on Enewetak Atoll either on platforms, barges, or underwater, 10 tests were conducted at Bikini Atoll, 2 tests near Johnson Atoll, and a high altitude test conducted about 100 kms west of Bikini Atoll.

Most nuclear tests conducted on Enewetak Atoll were detonated in the northern reaches of the atoll and produced highly localized fallout contamination of neighboring islands and the atoll lagoon. As a consequence, the northern islands on Enewetak received significantly higher levels of fallout contamination containing a range of fission products, activation products, and unfissioned nuclear fuel. By the time the test moratorium came into effect on 31 October of 1958, the United States had conducted a total of 42 nuclear tests on Enewetak Atoll.

Post Testing Era and Initial Cleanup Activities

Enewetak Atoll continued to be used for defense programs until the start of a cleanup and rehabilitation program in 1977. There were five feasible approaches considered by the Defense Nuclear Agency (NDA, 1981) for cleanup of Enewetak Atoll. The final plan called for (1) removing all radioactive and non-radioactive debris (equipment, concrete, scrap metal, etc.), (2) removing all soil that exceeded 14.8 Bq (400 pCi) of plutonium per gram of soil, (3) removing or amending soil between 1.48 and 14.8 Bq (40 and 400 pCi) of plutonium per gram of soil, determined on a case-by-case basis depending on ultimate land-use, and 4) disposing and stabilizing all this accumulated radioactive waste into a crater on Runit Island and capping it with a concrete dome.

Approximately 4,000 U.S. servicemen assisted in the cleanup operations, with 6 lives lost in accidents, in what became known as the Enewetak Radiological Support Project (DOE, 1982). A estimated total of 73,000 cubic meters of surface soil across 6 different islands on Enewetak Atoll was recovered by scapping and deposited in Cactus crater on Runit Island. The Nevada Operations Office of the Department of Energy was responsible for certification of radiological conditions of each island upon completion of the project.

The Operations Office also developed several large databases to document radiological conditions before and after the cleanup operations, and to provide data to update available dose assessments. The Enewetak cleanup program was largely focued on the removal and containment of plutonium along with other heavy radioactive elements. However, even during this early period of cleanup and rehabilitation, the adequacy of cleanup of the northern islands on Enewetak was brought into question because predictive dose assessments showed that ingestion of cesium-137 and other fission products from consumption of locally grown terrestrial foods was the most significant route for human exposure to residual fallout contamination on atolls affected by the nuclear test program

The people of Enewetak remained on Ujelang Atoll until resettlement of Enewetak Island in 1980. Between 1980 and 1997, the resettled population was periodically monitored for internally deposited radionuclides by scientists from the Brookhaven National Laboratory using whole body counting and plutonium urinalysis (Sun et al., 1992; 1995; 1997a; 1997b). More recently, the Department of Energy agreed to design and construct a radiological laboratory on Enewetak Island, and help develop the necessary local resources and technical expertise to maintain and operate the facility on a permanent basis.

This cooperative effort was formalized in a Memorandum of Understanding signed by the U.S. Department of Energy, the Republic of the Marshall Islands, and the Enewetak/Ujelang Local Atoll Government in August of 2000 (MOU, 2000). Construction on the Enewetak Radiological Laboratory was completed in May of 2001. The laboratory facility incorporates both a permanent whole body counting system, to assess radiation doses from internally deposited cesium-137, and clean living space for collecting in-vitro bioassay samples. Scientists from the Lawrence Livermore National Laboratory continue to support the operation of the facility and are responsible for systems maintenance, training, and quality assurance.

Rongelap Atoll

People and Events on Rongelap Atoll

On On March 1, 1954, the United States conducted a nuclear test on Bikini Atoll in the northern Marshall Islands code named Bravo that led to widespread fallout contamination over inhabited islands of Rongelap, Ailinginae, and Utrōk Atolls. Prior to Bravo, little consideration was given to the potential health and ecological impacts of fallout contamination beyond the immediate vicinity of the test sites. A total of 64 people living on Rongelap Atoll (including people residing on Ailinginae Atoll at the time of the blast) received significant exposure to "fresh" radioactive fallout and had to be evacuated to Kwajalein Atoll for medical treatment.

The Rongelap community spent the next 3 years living on Ejit Island (Majuro Atoll) before returning home to Rongelap in June 1957. However, growing concerns about possible long-term health effects associated with exposure to residual fallout contamination on the island prompted residents to relocate again to a new temporary home on Mejatto Island on Kwajalein Atoll in 1985. The people of Rongelap are still resident on Mejatto today although parts of the community have split off to live on Ebeye Island (Kwajalein Atoll) and Majuro Atoll.

The Rongelap community has always expressed a strong desire to return to their ancestral homeland. Through the Rongelap Resettlement Act, the United States Congress approved and continued a 1996 resettlement agreement between the United States and the Rongelap Atoll Local Government, and extended distribution authority for 10 years to advance resettlement. As a part of the 1996 resettlement agreement, a Phase I resettlement program was initiated in 1998. The United States Department of Energy, the Rongelap Atoll Local Government, and the Republic of the Marshall Islands have since signed a Memorandum of Understanding (MOU, 1999) outlining shared provisions in support of resettlement. Under this agreement, scientists from the Lawrence Livermore National Laboratory were tasked with developing individual radiation protection monitoring programs for resettlement workers and to verify the effects of the remedial actions.

Resettlement of Rongelap Atoll

Phase I resettlement of Rongelap Island is nearing completion. Rongelap Island now boasts a host of modern-day facilities including electrical power, a freshwater supply, a modern field station, paved runway, a number of bungalows for accomodating tourists and other visitors to the island, a whole-body counting facility and adjoining health physics laboratory, and a large concrete pier.

The remedial actions adopted under the Rongelap Resettlement Program are based on recommendations provided by scientists from the Lawrence Livermore National Laboratory. The remediation technique being employed is referred to as the combined option and involves replacing contaminated surface soil in the community-village area, where people spend most of their time, with a layer of clean crushed coral fill and adding potassium chloride fertilizer to the surrounding agricultural fields.

Limited soil removal and addition of coral fill reduces external exposure to gamma/beta radiation as well as inhalation exposure to radioactive contamination in the air that people breathe. The addition of potassium fertilizer to the agricultural areas competitively blocks cesium-137 uptake into plants, especially into the fruits of the main subsistence crops such as coconuts. It is expected that addition of potassium fertilizer on Rongelap Island will reduce the ingestion dose from cesium-137 to about 30 % of the pretreatment level and, at the same time, help support plant growth and increase the productivity of plants (see related information under Bikini Atoll).

After living in exile for nearly 2 decades, the prospect that the people of Rongelap will soon return to their ancestral homeland is an important milestone in the history of the Marshall Islands Program. Moreover, the Rongelap resettlement program is among the first in which a local government has engaged the United States Department of Energy to develop shared provisions to monitor the return of the population.

Reducing External Exposure

Contaminated soil around the proposed community center on Rongelap Island has been replaced with a layer of clean crushed coral to reduce external exposure to cesium-137 and other sources of penetrating radiation present in the underlying soil. The initial phase of this work was completed in March 2001. A detailed in-situ gamma monitoring survey of the entire community area was conducted in May 2001.

Additional in-situ gamma surveys were carried out in 2006 to assess external gamma exposure rates around homesites. The results of these studies clearly show that the combination of limited soil removal and addition of crushed coral fill is very effective in reducing external gamma exposure rates. The clean surface layer of coral also has the added benefit of reducing potential exposures from inhalation and ingestion of plutonium and other long-lived radionuclides present in the soils. View full report: UCRL-ID-143680-Pt-1 (Hamilton et al., 2001).

Utrok Atoll

People and Events on Utrōk Atoll

Utrōk Atoll is located about 500 kilometers east of Bikini Atoll. The atoll experienced significant radioactive fallout deposition from atmospheric nuclear weapons tests conducted in the Northern Marshall Islands during the 1950s. The most significant contaminating event on Utrōk Atoll was the Bravo test conducted at Bikini Atoll on March 1, 1954. The 167 residents (including 8 in utero) living on Utrōk Atoll at the time of the blast received significant external and internal exposures to fresh fallout contamination before being evacuated to Kwajalein Atoll on March 3, 1954. The Utrōk community returned to their home atoll 3 months later but continues to seek assurances from the United States Government that the atoll is safe for habitation.

The U.S. Department of Energy originally assigned responsibility for the internal dosimetry program on Utrōk Atoll to the Brookhaven National Laboratory. Through the 1990s scientists from Brookhaven conducted periodic whole body counting missions to the Marshall Islands to determine the body burdens of gamma-emitting radionuclides, such as cesium-137, cobolt-60, and potassium-40 in Marshallese from Bikini, Enewetak, Rongelap and Utrôk Atolls (Sun et al., 1997c). More recently, the U.S.

Department of Energy has developed a series of initiatives to address long-term radiological needs in the Marshall Islands. Under a working agreement between the Utrōk Atoll Local Government, the Republic of the Marshall Islands and the U.S.

Department of Energy (MOU, 2002), a permanent whole body counting system was established on Majuro Island (Majuro Atoll) during May 2003. With the cooperation of the Utrōk Atoll Local Government this facility also serves the general public, especially for those residents and visitors who return from the northern Marshall Islands who are concerned about being exposed to residual fallout contamination in the environment.

Under supervision from scientists from the Lawrence Livermore National Laboratory, the Utrōk Whole Body Counting Facility on Majuro is maintained and operated by Marshallese technicians. It is expected that Utrōk Atoll residents will be able to receive whole body counts during scheduled visits to Majuro under the routine medical surveillance program or on occasional outings.

Majuro is the capital city of the Republic of the Marshall Islands and is the main hub for the local airline.

Historical Data

Today, exposure to residual fallout contamination on Utrōk Atoll represents only a small fraction of the dose that people receive from natural sources of background radiation in the Marshall Islands. The radiological dose delivered to inhabitants living on Utrōk Atoll from residual fallout contamination in the environment is dominated by the external exposure and ingestion of cesium-137 (and to a lesser extent, strontium-90) contained in locally grown food crop products such as coconut, breadfruit and Pandanus.

According to Robison et al., (1999), the estimated population average maximum annual effective dose on Utrōk Atoll, based on a mixed diet containing imported foods, is less than 0.04 mSv (4 mrem) per year and has no consequence on the health of the population. Moreover, the predictive dose assessments based on environmental data and dietary models developed by the Lawrence Livermore National Laboratory appear to be in excellent agreement with estimates based on whole body counting (Robison and Sun, 1997).

Justification for establishing a permanent whole body counting system on Majuro Atoll for use by the Utrōk community comes from renewed concerns about high-end doses to maximal exposed individuals on Utrōk Atoll and that the associated health risk may exceed current guidelines adopted by the Marshall Islands Nuclear Claims Tribunal for cleanup of radioactively contaminated sites.

Such high-end individual doses in the Utrōk population have not been clearly demonstrated but the potential does exist for members of the population to binge on a local foods only diet or eat more foods containing higher than average radionuclide concentrations, e.g., coconut crab. Consequently, a permanent whole body counting program on Utrōk Atoll will provide a basis for conducting more accurate dose assessments across the entire population, and yield more detailed dosimetric data according to gender, age-group and seasonality.

Justification for intervention could then be made on the presumption that high-end doses are reasonably achievable, and that the risks from radiation exposures can reduced by means of remedial actions taking into account the relative cost:benefits as well as social and economic factors.

Nuclear Lawsuit

Marshall Islands Nuclear Lawsuit Reopens Old Wounds

The Republic of the Marshall Islands, the tiny collection of Pacific Ocean atolls and a former nuclear testing ground for the United States, is taking on the U.S. and eight other nuclear-armed nations with a set of lawsuits, claiming that the countries have failed to move towards disarmament and a world without nuclear weapons.

But they aren't seeking monetary compensation millions of dollars, along with a series of medical programs and cleanup operations, have been provided by the U.S. since they detonated dozens of nuclear and atomic bombs over the islands. How helpful the money has been remains a controversial topic.

Instead, the David and Goliath lawsuit claims the U.S. and its nuclear counterparts has failed to comply with the 44-year-old Nuclear Non-Proliferation treaty, which seeks to eliminate the international cadre of nuclear weapons, and promote the peaceful use of nuclear power. Filed on April 24, 2014, it seeks peace and adopts the line: If not us, who? If not now, when?

The Marshall Islands lawsuit, which was filed in U.S. Federal Court in San Francisco and in the International Court of Justice, claims that the U.S. has failed its duties under Article VI of the Treaty as it continues to modernize its weapons with no prospect of disarmament and is failing to pursue good faith negotiations, as set out in the treaty.

"[The Republic of the Marshall Islands] is not seeking financial compensation at all in these lawsuits. Their legal teams are working pro bono," Rick Wayman, director of programs at the Nuclear Age Peace Foundation, told *Newsweek* via email. "In these lawsuits, the RMI is seeking declaratory judgment of breach of nuclear disarmament obligations and an order that the nuclear-armed states initiate negotiations in good faith for an end to the nuclear arms race and to nuclear disarmament."

On July 21, the U.S. filed a motion to dismiss the lawsuit, in which President Barack Obama is named, along with the Departments of Defence and Energy and their secretaries and the National Nuclear Security Administration. The U.S. argues that their reasons for breaching their obligations are "justifiable, and not subject to the court's jurisdiction," according to a **statement** from the Nuclear Age Peace Foundation.

If not now, then when?

While the lawsuit focuses on the failure of the nine nations the U.S., U.K. Russia, China, France, Israel, India, Pakistan, and North Korea to dismantle their arsenals and eliminate of nuclear weapons, the effects on the islands and Marshallese people are long-lasting and likely to continue for generations.

The reaction to the lawsuit from the countries targeted has been mixed, ranging from Israeli doubts over the legal standing of litigation to outright surprise from Russia. France, a nation that is party to the Nuclear Non-Proliferation treaty and has less than 300 warheads and no reserve arms, is making repeated assertions that it is actively working to reduce its arsenal.

It's worth noting that Israel, Pakistan, India, and North Korea (although it used to be) are not covered by the Nuclear Non-Proliferation treaty but are bound by international law.

"The claim against Israel is different from the claim against the U.S. The claim against Israel is based on customary international law with regard to nuclear disarmament," Laurie Ashton, lead counsel for the Marshall Islands, said.

The people of the Marshall Islands have been living with the consequences of nuclear testing for decades, including forced resettlement and uninhabitable land, so why file the lawsuit now? It's a question that many, including the U.S., have asked.

Ashton says it's a question of "if not now, when?" Even though Obama "gave some hopeful words" in a speech in Prague in 2009, in which he lamented America's commitment to a world without nuclear weapons, Ashton points out that the U.S. have started new nuclear modernization plans. In 2014, there has been an 11.6 percent increase in funding for the U.S. nuclear stockpile over 2013, according to the Arms Control Association.

The U.S. continues to provide compensation to the islands, although questions swirl around the value of the money and the conditions of compensation. As part of the Compact of Free Association, an agreement between the Marshall Islands and the United States that went into effect in 1986, a $150 million compensation trust fund was agreed upon as the "'full and final settlement' of legal claims against the U.S. government." A provision prohibited Marshall Islanders from "seeking future redress in U.S. courts" and required the dismissing of all current court cases, according to the U.S. embassy of the Marshall Islands.

The $150 million figure "was completely pulled out of the air. There was no actual basis for it," Graham told *Newsweek* from the capital atoll of Majuro. By the time he retired in 2009, Graham said the settlement fund had been virtually exhausted and that while compensation could still be granted, paper awards instead of actual money was all there was left to give.

The U.S. government estimates that the total cost of compensation and assistance to the Marshall Islands is $531 million, including clean-up costs, according to a 2005 report by the Federation of American Scientists, while the U.S. government puts the amount awarded under the Marshall Islands Nuclear Claims Tribunal at $2 billion.

But what constitutes "real compensation" is a lingering question, Graham says. Only partial clean-up of certain atolls like Enewetak, where 43 atomic and nuclear bombs were detonated, mean that some islands remain off-limits. And the exhaustion of the $150 million Nuclear Claims Fund means that some people who have contracted cancers brought on or were directly caused by radiation will not be awarded any personal injury compensation, he says.

Again, the lawsuit doesn't seek monetary compensation. But it shows the damage done by radioactive waste on a nation. Graham spoke of the feeling on the islands: some Islanders are "apathetic" because no additional compensation is available, while others are wary about what the lawsuit will do to the nation's relationship with the U.S. The lawsuit shows that some are still angry at America.

"They maintain that this is damaging to the relationship the Marshall Islands has with the U.S. government, and possibly with other governments, including Israel, which has provided assistance to the Marshall Islands from time to time," Graham said.

"We didn't do this to annoy the United States and we certainly don't want to jeopardize the relationship or damage the relationship in any way. This is more in the nature of a friend or a younger brother saying, 'Hey, come on, do what's right here'," Graham said.

Cultural catastrophe

What the lawsuit can do, its advocates hope, is give voice and attention to a cultural catastrophe that nuclear testing wrought over the islands. In 2012, Calin Georgescu, then-United Nations special rapporteur on human rights and toxic waste, visited the islands. He reported that a solution to the nuclear legacy of the Marshall Islands still hadn't been found and that dislocation of people from their islands meant that "many communities 'feel like 'nomads' in their own country'."

And while the immediate effects of radiation like skin burns, vomiting, and hair loss later morphed into a biological legacy of reproductive issues and cancers laying dormant for decades, Palafox says that the absence of wellness and well-being, lingering posttraumatic stress disorders and the cultural damage and contamination of the environment are equally important.

“One of the things about the indigenous culture of the Marshall Island is they are very land-tied. All your powers… are tied to how much land you have, your relationship to the people,” Palafox said. “What the nuclear testing did on several of the atolls is when you contaminate the land, and you move the people to ‘safer islands,’ the whole cultural structure changes.”

“It had a very big effect on how the traditions passed to the next generation,” Palafox said. Women Palafox worked with constantly expressed worry that genetic disorders would be passed down to their children. Many atolls, each with their unique purpose, whether it be a burial ground for chiefs or a habit for birds, remain uninhabitable. Cultural behaviors were altered tremendously, Palafox said.

The Marshall Islands have until August 21 to file a response to the U.S Motion to Dismiss and the U.S. is due to respond to that by September 8. A hearing is scheduled for Sept. 12.

"Moreover, the NPT itself is not a lightswitch to be turned on and off at convenience -- States must be held to full accountability for violations of the Treaty or in abusing withdrawal provisions -- a matter of concern for every nation, and the wider global community that defines us all," Tony de Brum, foreign minister of the Marshall Islands, said at the United Nations in April

French-Polynesia opposition getting involve in Law pursuit

French-Polynesia's opposition party has announced that its future government will join the Marshall Islands in its International Court of Justice litigation against the nine nuclear states for breaching the Nuclear Non-Proliferation Treaty (NPT). Signing the treaty in 1968, the P5 states have committed to observing Article VI of the NPT, under which they are obliged "to pursue negotiations in good faith on effective measures relating to cessation of the nuclear arms race at an early date and to nuclear disarmament."

In April 2014, the Marshall Islands brought claims against the nine nuclear weapon states of the United States, the United Kingdom, China, Russia, France, Israel, North Korea, India and Pakistan, arguing that their continued possession of nuclear weapons amounts to a breach of the NPT.

For Marshall Islanders, this activism was borne out of tragic first-hand experience. From 1946 until 1954 the United States used the Marshall Islands as a nuclear testing ground. To this end, the indigenous inhabitants of the Pikinni, Roñļap, Wōtto, and Ānewetak atolls were forcibly resettled on outer island chains throughout 1946. As part of Operation Crossroads, two Hiroshima-size atomic tests were carried out, decimating Pikinni. These would be the first of a further 65 nuclear weapon trials within a twelve-year period.

Shielded from international attention, the U.S. carried out this testing with virtually no UN supervision. The experiment wreaked havoc on fragile Pacific ecosystems and, in some cases, changed the basic geography of the Marshall Islands forever. One such instance was the Mike test of November 1952, a bomb 750 times larger than the one dropped over Hiroshima, which vaporized the entire island of Elugelabl.

1954 saw the biggest test yet, with the Castle Bravo bomb – approximately 1300 times larger than Hiroshima – dropped over Pikinni on March 1, a date remembered by the Roñļap islanders as the “day of two suns.” It created a fireball four miles in diameter, vaporizing two islands and spreading a fallout zone of thick radioactive ash for 7000 square miles.

By 1963 the first cases of thyroid tumors, birth deformities, and severe growth retardation were diagnosed. The health and environmental impacts have no precedent in history; the magnitude of testing within the 1946-1954 period was the rough equivalent of 1.6 Hiroshima bombs detonating daily for twelve years.

Meanwhile, the indigenous peoples of French Polynesia have their own horrific stories to tell. Between 1960 and 1996, 196 weapons tests were carried out in French Polynesia by the French government, with recently de-classified documents revealing that during one 1974 test, Tahiti was exposed to 500 times the maximum allowed levels of plutonium fallout. Research has found a clear increase in thyroid cancer rates among people living within 1300 km of the tests, with one veterans' organization reporting thyroid tumors among some 85 percent of their members who had been exposed to testing.

For many activists, the struggle for justice for the survivors of the Mururoa atoll testing has become conflated with the ongoing decolonization process, begun three years ago when the French government agreed to allow the French Polynesian territories to be re-classified as a non-self-governing territory. Moreover, the declaration of support for the Marshall Islands case by the Tavini Huiraatira opposition party this week comes on the heels of the 2014 decision by the Assembly of French Polynesia to seek nearly $1 billion in compensation from the French government for the decades of testing.

The application contends that stockpiling weapons, improving nuclear technology and weaponizing nuclear material are violations of legal obligations owed by the nine states to the international community. The Marshall Islands further argues that these norms have crystallized into customary international law, obliging all states to pursue disarmament.

Failure to observe this commitment is, however, unlikely to amount to a breach of international law. One reason for this is that out of the nine states, only India, Pakistan and the United Kingdom have actually submitted to the compulsory jurisdiction of the ICJ under Article 36(2) of the court's statute. This means that the other six states must consent by special agreement to the court's jurisdiction in this particular case in order for the matter to progress, which is highly unlikely.

As the NPT heads into its 46th anniversary in 2016, prevailing state practice towards its obligations indicates that disarmament has a long way to go. With a total of 15,700 warheads in the world today, it will be some time before it can be said that the obligations contained within Article VI are binding on all states as a matter of customary law.

As a tiny Pacific country, the Marshall Islands are clearly outclassed in the ICJ, and are regarded by many as fighting a losing battle. Whilst Tavini Huiraatira has hinted that further Pacific Island countries will be joining the litigation in the near future, growing support is unlikely to have any effect on the solidifying NPT norms. Nevertheless, the litigation represents an important opportunity for these nations to tell their stories, to agitate for renewed commitment from signatory states, and to inject further impetus into the global disarmament movement.

Nuclear weapons and climate change

A double whammy for the Marshall Islands

On May 30, an unarmed intercontinental ballistic missile launched from the Kwajalein Atoll military base in the Republic of the Marshall Islands collided with an interceptor launched from California's Vandenberg Air Force Base. Planned for years, the shootdown was the first live-fire test against an ICBM-class target of a system designed to defend the United States against enemy missiles. Although many areas of the Marshall Islands are threatened by radioactive contamination from US nuclear testing during the Cold War and, more recently, by rising sea levels in a warming climate the

US military continues to use the Kwajalein Atoll for missile testing and other operations.

The United States is modernizing facilities in the Marshalls, just as it did during the age of nuclear testing. But these steps, seen largely as successes by the Pentagon and the American public, leave behind casualties. For the Marshallese, casualties came in the form of deadly radiation and displacement during the period when America was testing the most powerful weapons the world had ever seen. Today, islanders are faced with an equally harrowing issue, one that is only worsened by the testing done in the 1940s and '50s: rising seas that are threatening to destroy low-lying homes.

Nuclear testing and climate change together constitute a double whammy that is again forcing the Marshallese out of their homes. Because nuclear testing destroyed or irradiated much of the highest ground in the

Marshalls, it left the Marshallese with ever-limited options of where to go to escape the flooding that now threatens their lifestyles.

A nation at risk. The United States began nuclear testing in the Marshall Islands in 1946, and the decades that followed saw fatal radiation, failed relocation, and the Atomic Energy Commission's denial that radiation was an issue for the Marshallese. After forced evacuations from the Bikini, Enewetak, and Rongelap Atolls the most populated original settlements the Marshallese moved to smaller, lower-elevation islands.

There they were poisoned by irradiated fish, starved, flooded, and finally told it was safe to go back. The Bikini islanders' attempt at resettlement failed: Not only did they suffer from ground-level external doses of radiation, but also from various forms of ingested nuclear particles. After settling again on Bikini, the islanders were forced to re-evacuate after they began to suffer many ill effects of radiation. Today, echoes of this radiation still remain on the atoll, making it unsafe for human habitation.

The Marshallese struggled for years to get compensation. Finally, in 1986, representatives of the United States and the Republic of the Marshall Islands signed the Compact of Free Association, and reparation was approved for nuclear testing damages. Since then, the United States has provided millions of dollars toward legal resettlement fees, illnesses related to radiation, and atoll rehabilitation. (The Nuclear Claims Tribunal established by the two governments has awarded additional compensation to the islanders, but the US Congress and courts have thus far turned down islanders' petitions for this funding.)

As part of the Compact of Free Association, the Marshallese are permitted to work and live in the United States without a visa. Many have taken advantage of this opportunity, settling primarily in Hawaii, Arkansas, and Oklahoma. For those left behind in the islands, it isn't all tropical beaches and sunshine.

They must work hard to keep the waters at bay. According to a Marshallese local, the islanders have no choice but to repeatedly put up makeshift flood walls. Despite these precautionary measures, rising seas constantly damage houses and other buildings. Also, salt water that is invading the soil and underground stores of fresh water is causing a lack of sufficient drinking water and healthy crops. So why don't all the islanders move to the United States?

The situation is different for each family, but lack of money generally isn’t the primary reason for the Marshallese to remain on their home islands. By moving to America and leaving behind centuries of ancestry, the Marshallese endanger their unique island culture. Many islanders do not wish to uproot their entire families or leave behind close friends who aren't making the move.

Nowhere to run. Would the Marshallese have to move if the United States had not used their islands for nuclear testing decades ago? Yes, eventually. But not so soon. Many geographical landmarks could provide safe haven today if not for nuclear testing. Bomb craters are an ever-present reminder of the damage including a two-kilometer-wide, 50-meter-deep crater left by the first hydrogen bomb, "Ivy Mike," and another crater covered by the "Cactus Dome," a leaking 9,290-square-meter concrete dome on Runit Island built to contain nuclear debris. The "Castle Bravo" bomb test blew a hole through Bikini Atoll, and "Ivy Mike" not only vaporized an entire island but also produced waves that cleared surrounding islands of vegetation. These areas are now uninhabitable.

Radiation still plagues the Marshall Islands. Bikini Atoll remains the most radioactive area of all at 184 millirems per year, nearly double the predetermined safe amount agreed upon by the United States and the Republic of the Marshall Islands. If not for this radiation, the Marshallese could be living on Bikini, an area with a greater land mass and higher elevation than other major atolls and islands in the Marshalls, such as the Kwajalein and Rongelap atolls. Instead, the Marshallese are trying to make a living on smaller, lower islands the very places under assault by rising seas. The highest point in the Marshall Islands is only 10 meters above sea level. Meanwhile, the *average* elevation in the United States is more than 54 times that height.

The National Academy of Science has deemed other areas, such as the remote parts of Rongelap, safe because they meet predetermined radiation safety standards. Despite this assurance, many Rongelap residents are deeply fearful of radiation. Their alternative is Ebeye, the "slum of the Pacific" and the second most populated city in the Marshalls, an island where crumbling buildings, a cramped population, and sewage-filled streets are apparently seen as a better alternative than living near former test sites.

Ignoring the plight of the Marshallese is not good foreign policy. The United States is among the top polluters and carbon dioxide emitters in the world, and the Marshall Islands are suffering from those emissions. Islanders say it feels like they are living underwater already, and that is predicted to be their reality by the end of the century. The Marshallese regard climate change as an existential threat. When will the United States realize that the threat is as serious as nuclear testing once was?

With US President Donald Trump pulling out of the Paris Accord and proposing massive cuts to the EPA budget, the White House will do further harm to the Marshallese. The United States has presented the islanders with a double whammy: first nuclear testing forced them out of their higher-elevation settlements and made them sick with radiation, and now the US lack of action on climate change is forcing them off their islands *again*.

An American solution. On average, global sea levels are rising at about an eighth of an inch per year although the rise varies from location to location, and some reportshave estimated a rise of 11 to 38 inches by 2100. In the past century, the sea has already risen 4 to 8 inches. If this rate of rise continues, a temporary solution could come in the form of US-funded sea walls. Ideally, one-meter-high concrete walls with a curved, efficient design would last about 30 years, given maintenance and the durability of the material.

This would be enough to cope with an estimated 3-inch sea level rise during that period, and to block most of an average weekly high tide, which can get up to 1.89 meters. If funds cannot be raised for the undoubtedly expensive concrete design, a lower-cost option can be considered: an earthen mound wall. While sea walls are not a long-term solution, they would at least buy some time for the Marshallese.

A common assumption is that the threat of flooding gives the Marshallese enough reason to move from their islands. However, the islanders need a bigger incentive before they will risk losing their culture. One solution would be to equip the most popular settlements such as those already in Hawaii, Arkansas, and Oklahoma with multicultural centers, giving former islanders the tools to preserve their unique culture and heritage. Many Marshallese traditions are rooted in the ocean: ancient methods of sea navigation and canoe-building, for example. Elders could host activities or lessons to teach younger generations old island traditions.

They could learn about art, music, and cooking practices and pass on these traditions to future generations. Funding for these centers could come from a nonprofit organization, or a parks and recreation budget. The United States cannot reverse what it has done when it comes to radiation and climate change, but it can fortify Marshallese borders and accept the islanders as its own.

Moving forward, it is crucial to remember the past. Finding similarities between earlier nuclear testing and today's climate change isn't difficult. Both situations are characterized by denial and a lack of action. Reversing the effects of a warming world isn't realistic, just as reversing the effects of radiation back then weren't realistic. America's apologies cannot turn back the clock, but efforts can be made to improve the islanders' daily lives with sea walls and cultural organizations. The United States can also do more to ameliorate climate change, the biggest threat to the Marshallese in the coming century.

Looking back at how the Marshallese were treated serves as a reminder that no country, no matter how powerful, should have the ability to decide another nation's existence. The United States must be prepared to do everything possible to prevent this tiny island nation from disappearing beneath the waves.

Radiation Doses and Cancer Risks in the Marshall Islands

Associated with Exposure to Radioactive Fallout from Bikini and Enewetak Nuclear Weapons Tests

Overview

The Marshall Islands atolls were administered by the United States as a United Nations Trust Territory from 1947 until 1986 when the Republic of the Marshall Islands was established as a sovereign nation in free association with the United States. Previous to those years, the Marshall Islands were administered by Japan under a League of Nations mandate, and were the site of many important battles of the Pacific during World War II. After World War II, the United States established the Pacific Proving Grounds for testing nuclear weapons.

From 1946 through 1958, 65 nuclear weapons tests, in seven series, were carried out by the United States at Bikini and Enewetak Atolls located at the northwestern end of the archipelago that makes up the Marshall Islands and one additional test was carried out 100 km to the west of Bikini. The total explosive yield of the 66 tests was approximately 100 Mt (equivalent to 100 million tons of trinitrotoluene or TNT), about 100 times the total yield of the atmospheric tests conducted at the Nevada Test Site.

Radioactive debris from the detonations, dispersed in the atmosphere, was generally blown by the predominantly easterly winds towards the open ocean west of the Marshall Islands, though various historical reports indicate that radioactive debris from a number of tests traveled in other directions. The radioactive debris generated by the tests that eventually fell to the ground is termed fallout and was the single source of the exposures of the Marshallese people described in this report.

According to our analysis, twenty of the 66 tests that were carried out in or near the Marshall Islands resulted in measurable fallout in the Marshall Islands. Of special significance was the largest test conducted in the Marshall Islands, code-named Castle Bravo, a 15-Mt thermonuclear device tested on 1 March 1954. As a result of unexpected wind shear conditions, heavy fallout of debris from Bravo on atolls east of the Bikini Atoll test site resulted in high radiation doses to the populations of nearby atolls.

While the populations of Bikini and Enewetak were relocated before the testing began, other populations were evacuated following the Bravo test. Within about two days following the detonation of the Bravo test and the unexpected fallout on atolls to the east, the resident populations of Rongelap (including some Rongelap residents temporarily present on Ailinginae) and Utrik, as well as American military weather observers on Rongerik, were evacuated to avert continued exposure, to be decontaminated, and to receive immediate medical care for conditions of acute exposures .

In the month after the Bravo test, ^{131}I, an important radionuclide in fallout, was measured in urine collected about two weeks after the Bravo event from adults exposed on Rongelap, Ailinginae, and Rongerik. Those measurement data have proved to be of significant value for reconstruction of internal dose for those groups. For example, Brookhaven National Laboratory used the activity measurements in urine as well as other data and assumptions to estimate internal thyroid dose for persons exposed on Rongelap, Ailinginae, and Utrik. Internal doses from long-lived radionuclides on Rongelap and Utrik also were estimated by using whole-body and bioassay data collected years after the Bravo test.

The U.S. Government through Brookhaven National Laboratory and other institutions has provided decades of medical care, health surveillance, and documentation of health effects among the highly exposed Marshallese from Rongelap/Ailinginae and Utrik, but only two epidemiologic studies have ever been conducted, one of benign thyroid disease and one of benign thyroid disease and thyroid cancer. To date, there has not been a broad epidemiologic study of the Marshallese to determine the total numbers of cancers and other serious illnesses resulting from exposure to radioactive fallout. Nor has there been reliable diagnoses and recording of cancers among the general Marshallese population over the years since the nuclear testing ended that would now permit comparing their cancer rates with rates at other locations worldwide.

In 2004, the Senate Committee on Energy and Natural Resources asked the National Cancer Institute (NCI) for its "expert opinion" on the estimated number of baseline cancers ‡ and radiation-related illnesses from nuclear weapons testing in the Republic of the Marshall Islands. The Division of Cancer Epidemiology and Genetics (DCEG) of the NCI was tasked with developing a response because of its robust research program in radiation epidemiology and many years of experience in reconstruction of fallout-related doses and in cancer risk estimation.

For that purpose, we developed unrefined estimates of radiation doses and numbers of radiation-induced cancers, based on: (1) 1954 measurements of ^{131}I in the urine of adults exposed on two atolls, Rongelap and Ailinginae, collected after the test Bravo in 1954; (2)

measurements made in 1957–1977 of the contents of ^{137}Cs and other radionuclides in the bodies of inhabitants of Rongelap and of Utrik who returned to their atolls in 1957 and 1954, respectively; and (3) measurements of total ^{137}Cs and plutonium in soil from each atoll obtained for all atolls from the Marshall Islands-sponsored radiological survey completed in 1994.

We combined those elements using a simple analytic approach to develop crude estimates of the number of cancers likely to be radiation-induced among those living in 1954. This was, to our knowledge, the first time radiation doses and numbers of radiation-induced cancers had been estimated in a systematic manner over the entirety of the territory of the Marshall Islands. Our unrefined estimates were generally conservative and were intended to avoid under-estimation of the number of cancers that might occur. These initial results were presented during joint hearings of the House of Representatives Committee on Resources and the Committee on International Relations in May 2005. Following these joint hearings, we improved the models and data analysis to derive more realistic estimates of external and internal radiation dose by year, atoll, and age, as well as improved estimates of cancer risks. Those estimates and the methods on which they are based are the subject of this Summary paper and its companion papers.

The purpose of this group of papers is to present, in the peer-reviewed literature, a summary of the most important data that are available and that are useful for dose reconstruction, a detailed analysis of fallout deposited on each of the atolls of the Marshall Islands from nuclear weapons tests at Bikini and Enewetak, improved estimates of radiation doses, and improved estimates of cancer risks resulting from exposure to the fallout.

These estimates are based on a much deeper analysis of the available data than in and on models developed especially for this study. Although numerous studies have been conducted over the past decades to monitor the Marshall Islands and people, to develop land remediation strategies, and to assess contemporary and possible future doses that might be received by inhabitants of certain atolls in the Marshall Islands, the focus was more often on radiological monitoring, and on the northern Marshall Islands in particular.

Many of those studies were chronicled in a special issue of Health Physics. The current study, however, is the first comprehensive effort to estimate the deposition of all the important radionuclides contributing to dose and to estimate the doses and associated cancer risks to the population of the Marshall Islands.

The present paper summarizes the purposes and methods of the overall study and the estimated radiation doses and related cancer risks, as well as presents data that are common to all of the above papers, including the nuclear tests, the radionuclides considered, and the population sizes and their movements during the testing period.

Scope of The Study

The overall purposes of this study were to derive an internally consistent set of radiation absorbed doses to Marshallese alive during and after the years of nuclear testing in the Marshall Islands and to provide a thorough description of methods used in the dose reconstruction, to estimate the number of cancers that had already occurred and that could be attributed to radiation exposure, and to estimate the number of radiation-related cancers yet to occur.

The dose and risk assessment includes all Marshallese population groups and takes into account the size of the population of each atoll community, the baseline cancer risks (i.e., cancers unrelated to fallout exposure), and all of the Bikini and Enewetak nuclear tests that resulted in fallout over the territory of the Marshall Islands.

As indicated in, we estimated that, of the 66 nuclear tests detonated in or near the Marshall Islands from 1946 through 1958, 20 tests deposited measurable fallout in the Marshall Islands excluding the atolls on which the test sites were located. These tests were: Yoke in 1948; Dog and Item in 1951; Mike and King in 1952; Bravo, Romeo, Koon, Union, Yankee, and Nectar in 1954; Zuni, Flathead, and Tewa in 1956; and Cactus, Fir, Koa, Maple, Redwood, and Cedar in 1958. The characteristics of these 20 tests are presented in. Each of these 20 tests was taken into account in the estimation of radiation doses and cancer risks.

There are 30 atolls and four separate reef islands in the Marshall Islands. Ground deposition densities were estimated for 63 radionuclides plus 239,240Pu for all the atolls and separate reef islands except the two atolls where the testing sites were located (Bikini and Enewetak). However, some of the atolls were not inhabited during all or part of the testing period either because they were historically used only for gathering food (Ailinginae, Bikar, Erikub, Jabat, Jemo Island, Knox, Taka, and Taongi) or because the residents were relocated for safety reasons (Bikini and Enewetak) or evacuated due to unexpected exposures (Ailinginae, Rongelap, Rongerik, Utrik).

Thus, radiation doses were estimated for 26 population groups, including the residents of the 23 atolls and islands that were inhabited during the years of nuclear testing, and three other groups: persons from Rongelap who were on Ailinginae at the time of the Bravo test, persons from Rongelap who were visiting the southern atolls at the time of the Bravo test, and U.S. military weather observers on Rongerik.

For the consideration of cancer risks, only the 25 Marshallese population groups were considered. Both the dose and cancer risk assessment explicitly included members of the six Marshallese groups that were relocated or evacuated during the testing period: (1) 64 persons evacuated from Rongelap Atoll after the Bravo test; (2) 18 persons from Rongelap evacuated from Ailinginae Atoll after the Bravo test; (3) 117 persons from Rongelap who were visiting the southern atolls at the time of the Bravo test; (4) 157 persons from Utrik Atoll evacuated after the Bravo test; and the populations that normally resided on (5) Enewetak Atoll and (6) Bikini Atoll but who had been relocated to Ujelang Atoll and Kili Island, respectively, before the testing program began. Population data for all atolls and reef islands of the Marshall Islands at various times including the years of the nuclear testing period are presented in. Information on the dates of evacuation and on the places of relocation is provided in.

Radiation absorbed doses to the thyroid, red bone marrow (RBM), stomach wall, and colon wall were estimated for members of the 25 Marshallese population groups by age group (children under 1 y, 1–2 y, 3–7 y, 8–12 y, 13–17 y, and adults) and for the U.S. military personnel on Rongerik. Those specific organs and tissues were selected because they are expected to give rise to the largest number of cancers for reasons noted below:

- ✓ The thyroid gland, far more than any other organ, concentrates radioiodine, which is amply produced by detonations of nuclear weapons;
- ✓ Irradiation of the blood-forming cells in the RBM was caused mainly by external exposure to gamma-emitting radionuclides but also by internal exposure to radiostrontiums, and

would be expected to have increased the risk of leukemia, which has shown an especially strong relationship with radiation exposure in many epidemiologic studies; and

- ✓ The stomach and colon walls can be highly exposed after ingestion of fallout because many of the radionuclides produced by nuclear fission are highly insoluble, even in the gastrointestinal tract, thereby irradiating the stomach and colon as they pass through it.

The skin was also a tissue potentially exposed to fallout radiation. Marshallese who received significant amounts of fallout directly onto their body, e.g., at Rongelap where skin "burns" were documented, would have received high skin doses primarily from beta particles emitted during radioactive decay.

In this analysis, we have not estimated the dose to skin or the number of skin cancers that might be produced as a consequence of exposure to fallout, primarily for two reasons: (1) there are no baseline non-melanoma skin cancer data reported by the Surveillance, Epidemiology and End Results (SEER) program and other U.S. cancer registries, and the baseline risks are an essential part of the calculation to estimate the number of cancers, and (2) the number of personal injury claims awarded by the Marshall Islands Nuclear Claims Tribunal indicates that, among the 2,046 awards made through June 2004, there were 72 awards for skin burns, but only one award for skin cancer.

Hence, it appears that, despite potentially high doses to the skin to at least a small subset of the Marshallese, there is little evidence that the risk of skin cancer is great among Marshallese.

Estimated doses were derived for "representative" persons, that is, for persons who could be described to have habits, lifestyles, diet, and anthropometric characteristics typical of the Marshall Islands population for their age and sex (except in the case of military servicemen on Rongerik). Doses were assessed on a yearly basis for exposures occurring from 1948, the year in which the first relevant test took place, to 1970, when the residual environmental contamination had reached negligible levels on most atolls. These estimated annual organ doses were necessary input data for the cancer risk calculations.

The estimated total radiation absorbed doses include three components: (1) doses from external irradiation emitted by fallout deposited on the ground; (2) doses from internal irradiation from acute radionuclide intakes immediately or soon after fallout after each test; and (3) doses due to internal irradiation from chronic (i.e., protracted) intakes of radionuclides resulting from the continuous presence of long-lived radionuclides in the environment.

Sixty-three radionuclides, listed in, were considered in the estimation of internal doses from acute intakes of fallout radionuclides from each test. Based on screening estimates, these 63 radionuclides were estimated to account for over 98% of the internal dose to any organ from acute intakes. In addition, five long-lived radionuclides (55Fe, 60Co, 65Zn, 90Sr, and 137Cs) were considered for the estimation of the internal doses from chronic intakes, including two radionuclides, 60Co and 65Zn, that were not considered in the calculation of the doses from acute intakes. Doses from acute and chronic intakes from cumulative deposition of 239+240Pu were also estimated.

Risks of radiation-induced leukemia and cancer of the thyroid, stomach, and colon, as well as all other cancer types combined, were assessed for the 25 Marshallese population groups on the basis of the estimated radiation doses. Two time periods were considered:

from 1948 through 2008 for the assessment of the radiation-induced cancers that have been expressed thus far, and from 2009 onwards for the prediction of cancers that remain to be expressed. For comparison purposes, the numbers of baseline cancers, that is, those unrelated to fallout exposure, are also reported.

Summary of Methods and Findings

A brief overview of methods of the study and a summary of the findings are presented here. Detailed information can be found in individual companion papers. Throughout this section and elsewhere, we discuss findings relative to four groups of atolls or communities. Within each group, resident populations were exposed to similar levels of fallout as a consequence of the dispersion patterns of the nuclear debris clouds.

The southern atoll group is well represented by Majuro, which is the national capital today and was home to about one-third of the population of the southern atolls in 1958, while the mid-latitude atolls are best represented by Kwajalein, which was home to about one-quarter of the total Marshall Islands population during the testing years. Our radiological findings for the southern atolls and mid-latitude atolls along with our radiological findings for the Utrik community and for the Rongelap Island community (both from the northern atolls) capture the range of exposures received by Marshallese at all atolls.

In the case of Utrik and Rongelap, we define the “community” to be those exposed to fallout from the Bravo test on Utrik and Rongelap, respectively, and who were evacuated after the Bravo test. Our findings illustrate the geographic pattern as well as provide atoll and atoll-group estimates of contamination, organ dose, and cancer risk as well as the dependence on age at exposure.

Fallout activity deposited on the ground

As discussed in, a complete review of various historical and contemporary deposition-related data, some available only in gray literature (e.g., government laboratory reports and internal agency and laboratory memoranda, supplemented by meteorological analyses) was used to make judgments regarding which tests deposited fallout in the Marshall Islands and to estimate fallout deposition density and fallout transit times, otherwise known as times-of-arrival (TOAs). In some instances, it was necessary to use the results of a well-established model of atmospheric transport and deposition to corroborate or contradict our initial assumptions on the occurrence of fallout on particular atolls after certain tests.

The various types of data reviewed for estimating deposition included measurements of 137Cs and other radionuclides in soil (both historical and contemporary), historical measurements of exposure rate following individual tests derived from aerial surveys, ground surveys and continuous-reading monitoring devices (strip-chart recorders), and historical measurements of beta activity collected on gummed film during the years of nuclear testing.

For each atoll, fallout TOAs and the estimated fractionation of fallout were used to estimate deposition density for 63 activation and fission products from each nuclear test, plus the cumulative deposition over all tests of 239+240Pu. Examples of deposition densities of 24 of these radionuclides are presented in.

The estimated total 137Cs activities deposited by all tests from this analysis, after appropriate decay to account for the effective decay rate (radiological plus weathering) in the Marshall Islands and a correction for global fallout from non-Marshall Islands tests, were compared with contemporary measurements of the total 137Cs activities remaining in the soil as measured by investigators in 1978 and in 1991–1993. This comparison was used to demonstrate the validity of our estimates of total 137Cs deposition density. Our atoll-specific cumulative 137Cs estimates were found to be in excellent agreement with contemporary measurements of 137Cs in soil.

Our estimates for the 137Cs deposition density and for the corresponding TOA at each atoll and for each of 20 individual tests are presented in tabular form by. Our best estimates of the cumulative 137Cs deposition density from all tests, with 90% uncertainty ranges, are presented in and the geographic pattern of total fallout deposition is illustrated in. The cumulative 137Cs deposition densities are much greater on northern atolls (e.g., Rongelap and Rongerik) than on mid-latitude atolls (e.g., Kwajalein) or southern atolls (e.g., Majuro), as can be noted, also provides estimates of deposition separately for southern and northern islands in Kwajalein Atoll and in Rongelap Atoll.

The deposition densities differed by about 20% between south and north islands of Kwajalein but more than three times between islands of south and north Rongelap Atoll, reflecting differences in deposition due either to the large size of the atoll (Kwajalein), or, in the case of Rongelap, to the position of the Bravo debris cloud trajectory relative to location of individual islands in the atoll.

The estimates of radionuclide deposition density, fractionation, and transit times reported in allowed estimations of both external and internal dose to representative persons as described in companion papers.

Radiation doses

As noted earlier, the estimated doses came from three sources of exposure: (1) external irradiation from fallout deposited on the ground; (2) internal irradiation from acute radionuclide intakes immediately or soon after deposition of fallout from each test; and (3) internal irradiation from chronic intakes of radionuclides resulting from the continuous presence of long-lived radionuclides in the environment.

External doses

The doses from external irradiation arose from gamma rays emitted during radioactive decay of the fallout radionuclides during the passage of the radioactive cloud or after deposition on the ground. Doses received during the passage of the radioactive cloud are generally insignificant compared to those delivered after deposition of fallout on the ground. Exposure during cloud passage was implicitly included by integration of the exposure rate from the initial time of fallout arrival rather than from the time when the exposure rate was at its peak.

The doses from external irradiation were estimated in three basic steps:

1. estimation of the outdoor exposure rates at 12 h after each test and of the variation in the exposure rates with time at each atoll after each test;
2. estimation of the annual exposure from 1948 through 1970 and of the total exposure from TOA to infinity, obtained by integrating the estimated exposure rates over time; and
3. estimation of the annual and cumulative absorbed doses to tissues and organs of the body by applying conversion factors from free-in-air (outdoor) exposure to tissue absorbed

dose and by assuming continuous residence on the atoll (with corrections for temporarily resettled populations).

The outdoor exposure rates at each atoll were assessed in one of two ways depending on whether reliable measurements of exposure rates were available for a particular nuclear test and atoll combination. If measurement data were available, they were assessed and a best estimate of the average exposure rate at 12 h post detonation (termed E12) on the atoll or reef island was made.

If no reliable exposure rate data were available to estimate E12 directly, then the assessment of E12 was derived from the estimates of 137Cs deposition densities and TOA provided in for each atoll and each test. The method relating the estimates of 137Cs deposition densities and TOA to E12 was developed by the Off-Site Radiation Exposure Review Project

(ORERP) for estimating external whole-body dose from fallout originating at the Nevada Test Site .

The annual and cumulative exposures derived from the estimate of E12 were estimated by using the variation with time of the exposure rate calculated by, but modified to take fractionation into account, where necessary, as well as the “weathering effect” which reflects the gradual decrease of the exposure rate caused by the migration of the deposited activity into deeper layers of soil.

The conversion factors from free-in-air (outdoor) exposure to tissue absorbed dose depend on the energy distribution of the gamma rays that are incident on the body and on the organ for which the dose is being estimated. However, for most of the fission and activation products that are created during a nuclear explosion, the gamma-ray energies resulting in external exposure are a few hundred keV or more and the variation in photon energy results in at most a few percent difference in dose per unit incident fluence for the various organs considered in this study. Thus, energy and organ dependence in dose conversion factors were not taken into consideration; a single conversion factor, 6.6 × 10–3 mGy per mR, was used for adults for all organs. However, the conversion factor does depend on the age of the person, or, more precisely, her or his body size and shape.

Thus, based on calculations using anthropomorphic phantoms that represented different ages, our calculated doses to adults from external irradiation were increased by 30% for children less than 3 y of age and by 20% for children 3 y of age through 14 y. While age and body size were important for the estimation of external dose to the organs considered, sex was not. Building shielding was estimated not to be important since houses at that time, made primarily out of palm fronds, did not provide any substantial reduction of gamma ray intensity.

Annual absorbed doses from external irradiation from all important tests were estimated for the time period from 1948 through 1970; that is, until the annual doses had decreased to very low levels in comparison to the peak values observed in 1954. These annual doses were estimated for the relocated populations and for the populations continuously resident on all inhabited atolls of the Marshall Islands in three age categories (infants, children, and adults). The doses reported for the relocated populations include, where appropriate, contributions from exposures received before evacuation, during the period of resettlement, and following return to the atoll of origin.

Annual doses to adults from external irradiation are presented in along with estimated uncertainties; the doses were highest during the years of atmospheric testing in the Marshall Islands, after which they decreased to values that were, in 1970, less than 0.1% of the peak values observed in 1954. Our best estimates of the total external doses (mGy) from all tests and of the 90% uncertainty ranges are presented in for representative adults of all 26 population groups. The geographic pattern of total external doses received is the same as for the deposition of 137Cs illustrated in and, as described, is much higher in the northern atolls than in the central and southern atolls.

Internal doses from acute intakes of radionuclides

The internal radiation doses resulting from acute intakes, defined as those that occurred during or soon after fallout deposition, were assumed to be primarily a consequence of ingesting radionuclides in, or on, debris particles that contaminated food surfaces, plates and eating utensils, the hands and face, and, to a lesser degree, drinking water. Internal doses from other pathways of exposure, in particular, inhalation, were much lower than those due to ingestion and have not been explicitly estimated in this assessment.

Fallout particles at northern atolls were typically large (tens to more than one hundred micrometers in diameter) resulting in generally low intakes by inhalation. Fallout deposited at southern atolls, even though generally composed of smaller sized particles, was often deposited with rainfall which significantly reduced the availability of the particles to be inhaled. Annual rainfall rates are three to four times greater in the southern atolls compared to the northern atolls.

The methods used in this study for estimating acute intakes of fallout radionuclides and resulting doses are based on: (1) the estimates of test-, atoll-, and radionuclide-specific deposition densities discussed in; (2) historical measurements of 131I in pooled samples of urine collected from adults about two weeks after the Bravo test and (3) assessment of appropriate values of gastrointestinal uptake for the radionuclides present in fallout particles.

The assessment of internal doses was composed of the following six steps: (1) estimation of the intake of 131I by populations on Rongelap, Ailinginae, and Rongerik, following the Bravo test using historical bioassay data; (2) estimation of the intake of 137Cs at the same three atolls based on the ratios of 137Cs to 131I calculated but corrected for fractionation; (3) estimation of the deposition density of 137Cs following each of 20 tests on all inhabited atolls; (4) estimation of the intake of 137Cs at all inhabited atolls assuming that the ratio of intake to deposition was the same at all atolls; (5) estimation of intakes of all radionuclides considered at all inhabited atolls following each nuclear test; and (6) estimation of annual and cumulative radiation absorbed doses to four organs (RBM, thyroid, stomach, colon) of representative persons for all relevant birth years.

Detailed information on the acute intakes and resulting doses, as well as the estimated uncertainty in these dose estimates, is presented in Simon et al. (2010). The population of the southern atolls had acute intakes estimated to be much smaller than those experienced by the more highly exposed Rongelap and Utrik populations. For example, adult Majuro residents had intakes of about 6% and 9% of the 131I and 137Cs (cumulative over all tests), respectively, of adult Utrik community members, and about 1%, and 2%, respectively, of the intakes of Rongelap community members exposed to Castle Bravo fallout on Rongelap Island.

Doses to the thyroid gland were much greater than those to the other organs and tissues, and were much greater for the Marshallese who resided on Rongelap and Utrik Atolls at the time of the Castle Bravo test than for the residents of any other atoll. The southern atolls, where about 73% of the population resided during the testing years, received the lowest organ doses.

The population of mid-latitude atolls (Kwajalein and others, home to about 23% of the total Marshall Islands population during the testing years, received organ doses that were about three times greater than at the southern atolls. The population of Utrik received doses intermediate in magnitude between the mid-latitude atolls and Rongelap, with thyroid doses about 35 times greater than the southern atolls. The Rongelap Island community received the highest doses, with thyroid doses about 350 to 400 times greater than those received in the southern atolls.

Internal doses from chronic intakes of radionuclides

Following the deposition of radionuclides on the ground, chronic (i.e., protracted) intakes took place at rates much lower than those due to the acute intakes. While both acute and chronic intakes were primarily a result of ingestion, the environmental transport processes leading to chronic intakes were substantially different from those that gave rise to acute intakes. Chronic intakes were primarily a function of the consumption of seafood and of locally grown terrestrial foodstuffs internally contaminated with long-lived radionuclides via root uptake and, to a lesser degree, inadvertent consumption of soil. A previous assessment showed that five radionuclides account for essentially all the internal dose from chronic intake: 55Fe, 60Co, 65Zn, 90Sr, and 137Cs.

The available historical whole-body counting and bioassay measurements were used as a basis to estimate the chronic intakes since a suitable dietary model covering the many years after the tests, when lifestyles became more westernized, does not exist. Those whole-body and bioassay measurements were made on the Rongelap and Utrik evacuees for years after they returned to their respective home atolls.

The Rongelap and Utrik populations, who were evacuated within about two days following the detonation of the Castle Bravo test on 1 March 1954, were returned to their home atolls in June 1957 and June 1954, respectively. During the first few weeks after their return and until the 1980's, a Brookhaven National Laboratory team regularly conducted measurements of whole-body activity of 137Cs, 60Co and 65Zn, as well as urinary concentrations of 90Sr. Measurements of 55Fe in blood were also performed, but only once.

The steps used to estimate the doses from chronic intakes of radionuclides were: (1) estimation of the chronic intakes by Rongelap and Utrik adult evacuees due to the Bravo test; (2) estimation of the chronic intakes resulting from the Bravo test by adults of all other atolls, based on the relative 137Cs deposition; (3) estimation of the chronic intakes by adults resulting from tests other than Bravo, again based on relative 137Cs deposition; (4) estimation of the chronic intakes by children; and (5) estimation of the doses from chronic intakes from all tests and all population groups using International Commission on Radiogical Protection recommended dose coefficients.

Detailed information on the estimation of chronic intakes and resulting doses is presented in. The doses from chronic intakes show the same geographical pattern as the doses resulting from acute intakes and 137Cs deposition. However, because of the absence of short-lived iodine isotopes which dominated the thyroid dose from the acute intakes, the thyroid doses from chronic intakes were not much greater than the doses to other organs and tissues.

Similar to the situation for acute intakes, only a few radionuclides contributed most of the organ absorbed dose. For all organs and for all four of the atoll and population groups discussed, 137Cs was either the first or second most important contributor to internal dose from chronic intakes. For the evacuated Rongelap Island community, 137Cs was the most important contributor to the chronic dose, whereas 65Zn was the largest contributor to dose for the residents of all other atolls.

The cumulative thyroid doses (mGy) to representative adults on each atoll from both acute and chronic intakes of radionuclides in fallout from all tests with 90% uncertainty ranges are presented in and have the same geographic pattern as for 137Cs deposition.

Comparison of doses by mode of exposure

Compares estimated cumulative internal doses to representative adults of four population groups as reported in Simon et al. (2010) with the external doses for those same population groups as reported in Bouville et al. (2010). As elsewhere in this paper and companion papers, those persons of adult age (>18 y) at the time of the first test with significant deposition (Yoke test, 1 May 1948) are considered as adults in this assessment. In addition, all dose estimates presented are best estimates based on an analysis of all available data.

With respect to the components of the internal dose, the dose from chronic intake exceeded the dose from acute intake for RBM and stomach wall, for all populations groups except the Rongelap Island community. For the Rongelap Island community, the acute doses for all organs exceeded the chronic doses. Because of the exposure to radioiodines in fallout, the absorbed dose to the thyroid gland from acute intakes exceeded the chronic dose to the thyroid, regardless of the population group. Acute doses to colon wall were also greater than the corresponding chronic doses for all four population groups.

With respect to the total internal dose relative to the external dose, external doses were much greater than the internal doses to RBM and stomach wall, regardless of the population group, but were comparable to the internal doses to the colon wall (greater by two-fold at the southern and mid-latitude atolls, and about one-half for the Utrik and Rongelap Island communities). Internal doses to the thyroid were significantly greater than external doses, regardless of the population group.

Total doses

Total (external plus internal) organ absorbed doses can be presented in various ways to demonstrate the spatial and time-dependence of exposures received across the Marshall Islands and the dependence on age at exposure. As discussed earlier, illustrates the groups of the atolls within the Marshall Islands with similar degrees of deposition. In parallel, presents population-weighted total doses to adults within each of the four geographic areas.

We found that our estimated total doses are relatively comparable within each of the four population groups: residents of southern atolls, residents of mid-latitude atolls, the Utrik community, and the Rongelap Island/Ailinginae/Rongerik evacuees. Here, as elsewhere in this paper and companion papers, demonstrate that adults in mid-latitude atolls received cumulative organ doses approximately four times as great as adults in the most southern atolls. Similarly, adults of the Utrik community received cumulative organ doses four to seven times as great as adults from the mid-latitude atolls. Adults among the Rongelap Island/Ailinginae/Rongerik evacuees received the largest cumulative doses, six to eight times as great as adults from Utrik.

Recognizing that the doses within each of the four areas can be represented by the doses to Majuro residents, Kwajalein residents, the Utrik community, and the Rongelap Island community, an illustration provides cumulative radiation doses (external plus internal) at those atolls for all birth years from 1930 to 1958. Those born in or before 1930 would be of adult age at the time of the first tests and would have received approximately equal doses regardless of the birth year.

We found that our estimates of total organ radiation absorbed doses (sum of external and internal) varied by year of birth. Persons who were adults at the beginning of the testing period (born in 1930 or earlier) received relatively low thyroid doses from the large tests in 1954 compared to those who were very young at the time of those tests . Among the four representative population groups, cumulative thyroid doses ranged from 33 mGy for adults who lived on

Majuro at the time of testing to as high as 23,000 mGy for infants on Rongelap Island at the time of the Bravo test.

The dose contributions from the six tests that resulted in the highest total doses (external plus internal) to adults of the four representative atolls and for the four organs and tissues that are considered (RBM, thyroid, stomach wall, and colon wall). Bravo was by far the most important contributor to the total dose for the Utrik and Rongelap Island communities, but less important than Yankee, Yoke, Koon, Romeo, and Flathead for the Kwajalein residents, and less important than Koon and Romeo for the Majuro residents.

For purposes of cancer risk projection, the annual organ doses are required. Annual doses were greatest in the years with large yield nuclear tests, i.e., 1954, 1956, and 1958. A figure shows the temporal pattern of total dose (external plus internal) received on an annual basis to the thyroid gland of children born in 1953 in each of four population groups (Majuro residents, Kwajalein residents, Utrik community, and Rongelap Island community). Children born in 1953 would have received the largest doses of any birth cohort.

Uncertainties of estimated doses

Estimated doses and the uncertainties associated with those estimates varied by location, fallout event, calendar year, and age at time of exposure. The precision of our dose reconstruction is better for exposures received on Rongelap and Ailinginae than on Utrik, primarily because of the availability of historical urine bioassay data and large amounts of environmental monitoring data (both historical and contemporary), and both are more reliable than the estimated doses for persons exposed on the mid-latitude and southern atolls.

Of the exposure pathways examined, determination of dose from external sources was the most direct and, therefore, the most precise. An analysis was conducted to evaluate the uncertainty in the annual doses from external irradiation for each year of testing. Annual and cumulative exposures were often estimated from historical measurements or from relatively simple conversions from fallout deposition density.

We determined that the uncertainty of doses from external irradiation could be characterized by lognormal distributions with geometric standard deviations (GSDs) of approximately 1.2 for exposure on Rongelap and Ailinginae, 1.5 for exposure on Utrik, and 1.8 for exposures on the other atolls. As can be seen, the overall GSDs were smallest for the communities where the greatest doses were received from the 1954 tests. Conversely, the GSDs were largest for communities with the lowest doses from the 1954 tests.

In comparison to estimates of external dose, estimates of dose from internal irradiation are substantally more uncertain. Based on an analysis accounting for uncertainties in the most relevant and sensitive parameters involved in the internal dose assessment, we found that the uncertainty of doses from internal irradiation could be characterized by lognormal distributions with GSDs of approximately 2.0 for exposure on Rongelap and Ailinginae, 2.5 for exposure on Utrik, and 3.0 for exposures on the other atolls. Doses from chronic intake of radionuclides result from a more complex exposure situation and are more uncertain than the doses from acute intakes. However, doses from chronic intakes were small and refinements to the estimation of the uncertainty associated with them would contribute little to the overall dose uncertainty.

Projected cancer risks

The annual doses from external irradiation and from internal irradiation that were estimated for the 25 Marshallese population groups according to birth year were combined with the population sizes and with age-dependent organ-specific risk coefficients to derive the corresponding cancer risk projections presented in Land et al. (2010). Risk estimates were presented in terms of the number of cancers by organ site projected to occur among Marshallese as a consequence of exposure to fallout from regional nuclear tests.

The cancer risks were based on an estimated population of 12,175 residents of the Marshall Islands born before 1948 and another 12,608 born in the years 1948 through 1970, giving a total potentially exposed population of 24,783. Projected lifetime numbers of baseline and radiation-related (excess) cancers are shown in an illustration table not shown here, by cancer type: leukemia, thyroid, stomach, and colon.

In addition, the numbers of "all other solid cancers" has been estimated using the colon dose as representative of the dose to most other organs and tissues of the body. The projected number of baseline (non-radiation related) cancers among the 24,783 Marshallese in all organs totals 10,600, while the projected number of excess (radiation-related) cancers is 170, including 65 that have yet to occur. In comparison to our 2004 estimates, which also are presented in a table, the numbers of projected radiation-related thyroid and colon cancers are much smaller as a result of a much more realistic dose assessment.

When the entire population of the Marshall Islands is considered, the estimated fraction of cancers that has occurred or will occur and that can be attributed to exposure to radioactive fallout, expressed as a percentage, is about 20% for thyroid and about 5% for leukemia. These percentages can be compared to all other cancers, for which the attributable fractions are on the order of 1%. The attributable fractions, as expected, were much higher among the most heavily exposed population groups.

A breakdown of the estimated number of cancer attributable to exposure to fallout radiation according to population group and time period, as well as estimation of the uncertainties in the projected number of cancers, is discussed in detail by Land et al. (2010). The attributable fractions (%) of all cancers from exposure to fallout radiation within each of the four atoll groups with 90% uncertainty ranges are presented in Table 11 and have the same geographic pattern as for 137Cs deposition. Because of the small numbers of projected cases on some atolls (resulting in highly uncertain estimates), the cancer risk projections are shown only for groups of atolls rather than for individual atolls.

IN CONCLUSIONS

The methods and findings described in this paper and the seven companion papers represent the most comprehensive retrospective evaluation ever conducted of exposure of Marshallese and the related cancer risks from regional nuclear testing. This effort, in response to a Congressional request, will provide information useful to U.S. Congressional committees as well as to health authorities both in the U.S. and in the Marshall Islands.

However, the methods are also illustrative of methods that may be useful in broader circumstances, some of which might occur in the future. Though nuclear testing in the atmosphere is not likely to be revived, nuclear detonations that would result in exposure of the public might occur in the future due to accidents or intentional actions in wartime or by terrorists. A number of important lessons can be derived from this analysis.

Here, we have confirmed that exposure to radioactive fallout, particularly soon after detonation of a large device, can result in high exposures and substantial increases in cancer risk. At distances of more than a few hundred kilometers, however, exposures and related cancer risks are likely to be highly diminished due to dilution of the radioactive debris in the atmosphere (depending on the meteorological conditions) and radioactive decay during transit.

Lifestyles that are dependent on storing and preparing food outdoors are particularly susceptible to transmitting radioactive contamination to man. Reconstruction of radiation doses many years after exposure can be an intensive effort and underscores the need for dependable data of various types. The amount of data necessary to make reliable estimates of radiation dose and cancer risks is significant and the collection of that information should not be overlooked following nuclear events, but should be, in fact, a high priority

CPSIA information can be obtained
at www.ICGtesting.com
Printed in the USA
LVHW090231200819
628263LV00002B/311/P